MINI COLOR SERIES 7511

M-113 in the 1990s (Part 1)

Text and Photos by Carl Schulze
Edited by James R. Hill
Illustrations by Hubert Cance

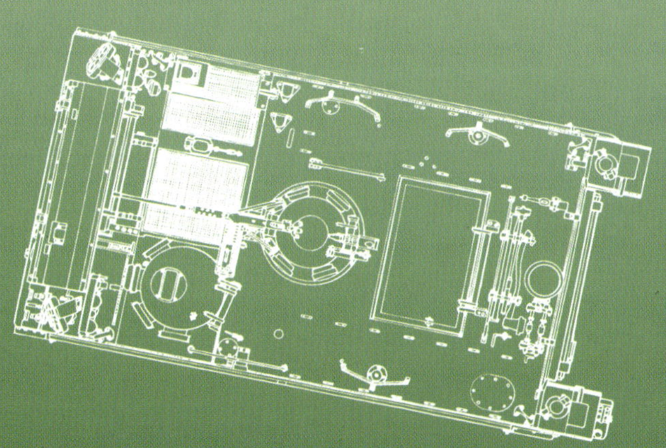

Editor: James R. Hill

by CONCORD PUBLICATIONS CO.
603-609 Castle Peak Road
Kong Nam Industrial Building
10/F, B1, Tsuen Wan
New Territories, Hong Kong
www.concord-publications.com

We welcome authors who can help expand our range of books. If you would like to submit material, please feel free to contact us.

We are always on the look-out for new, unpublished photos for this series. If you have photos or slides or information you feel may be useful to future volumes, please send them to us for possible future publication. Full photo credits will be given upon publication.

ISBN 962-361-677-5
printed in Hong Kong

The M113 in the 1990s (Part I)

The M113 was originally designed as an Armored Personal Carrier (APC) in the late 1950s. Since the time the series production of the vehicle started in 1960, more than 80,000 M113 and M113-based vehicles have been built. The M113's evolution has seen three major upgrading programs, and the end of its service life is not yet known. By 1990, most nations originally fielding the M113 as an APC had abandoned the vehicle in this role and either replaced the M113 with a new combat vehicle or with an M113-based vehicle such as the Armored Infantry Fighting Vehicle (AIFV). However, most of these nations still fielded the M113 during the 1990s in countless variants such as command post vehicles, mortar carriers, and anti-tank weapon platforms, etc. Manufactured by United Defense LP Ground System Division, Santa Clara, California, the M113 is used in more than 40 major variants and is in service with more than 50 nations. This book focuses on the M113 family of vehicles used in the 1990s. We decided to deal with the subject nation by nation rather than by variants.

Developmental History

In 1956, the U.S. Army Ordnance Tank-Automotive Command initiated a design program in order to develop a new lightweight, armored, amphibious personnel carrier with the capability to be air transported for the U.S. Army's armored and infantry units. The new vehicle was to replace the M59 and M75 armored personnel carriers then in service, which had high production costs, weighed too much and could not follow the latest U.S. main battle tank, the M48, across the battlefield. Several concepts were studied. One was called T113 and featured an aluminum-constructed hull, and another, called T117, featured a steel-constructed hull. Both concepts reached prototype status and were consequently built for trials by the successful contender FMC (Food Machinery Company).

After trials, the aluminum T113 was chosen because of its higher degree of ballistic protection, durability and cost effectiveness. Development continued with a modified prototype, the T113E1, which later was standardized by the U.S. Army as Carrier, Personnel, Full Tracked Armored, M113. Series production started in the FMC facility at San Jose, California, and the first M113 rolled out in June 1960. The first U.S. Army order was for 900 vehicles. Early members of the M113 vehicle family were the M577 command post vehicle and the M106 107mm mortar carrier. Although the first vehicles were delivered to the U.S. Army, the M113 saw its first operational service with the Army of the Republic of Vietnam (ARVN) in the Vietnam War, where it was deployed from 1962 onwards.

M113A1 Series

While the M113 production was already underway, the development of the vehicle continued. In 1962, the first M113A1 prototype was manufactured, which was called T113E2. The vehicle was fitted with a General Motors diesel engine instead of the original Chrysler V8 75M petrol engine. The diesel engine was incorporated into the M113 APC in order to extend the operational range and minimize the risk of engine fires, which happened regularly with the petrol engine. Together with the General Motors 6V 53 water-cooled, two-stroke, diesel engine, the Allison TX-100-1 transmission was installed in the M113A1. By 1964, when the M113A1 entered production, FMC had already produced 10,000 vehicles of the original M113. The M113 family members were now also delivered with the A1 modifications. The M577A1 and M106A1 also were joined by new family members such as the M125A1 81mm mortar carrier, the M741, which is the chassis of the M163 "Vulcan" air defense vehicle, the M579 Fitters Vehicle and the M113 Light Assault Bridge. In addition, a series of vehicles is based on the M113A1 chassis, which includes vehicles such as the M548 tracked cargo carrier, the M730 Chaparral surface-to-air missile system, the M752 Lance missile system launcher, and its M688 transporter/loader vehicle.

M113A2 Series

In 1978, the U.S. Army decided to launch a production improvement program in order to enhance the reliability and performance of its existing fleet of over 18,000 M113 and M113A1. The decision was made after successful trials with several prototype vehicles, which featured an improved engine cooling system and an improved suspension system. The suspension improvements consisted of high-strength steel torsion bars, improved shock absorbers and a stronger raised rear idler assembly. The improvements of the engine cooling system included a new radiator, fan and surge tank, as well as a redesign of the system in which several elements were moved to different positions. With the newly designed engine cooling system, the fresh outside air was now sucked through the radiator while in the past it had first been sucked in over the engine and then pressed to the outside through the radiator. The new cooling system increased cooling efficiency, minimized oil, film, and dust build-up on the radiator and, in addition, achieved a negative engine compartment pressure, which minimized the possibility that toxic engine fumes might enter the troop compartment. With the M113A2, the rear-mounted external armored fuel tanks (EFT) were available as a production option.

Since the time production began on the M113A2 in 1979, many non-U.S. orders have required the fitting of rear-mounted external armored fuel tanks, which provide more internal space for the crew and their equipment inside the vehicle's hull and reduce the risk of fire inside the vehicle. The M113 vehicle family with the M113A2 upgrade again welcomed new members, among them the M1059 Smoke Generating System, the M981 Fire Support Team Vehicle (FISTV) and the M901 Improved TOW Vehicle (ITV).

M113A3 Series

Between 1987 and 2001, the U.S. Army modernized their entire M113 vehicle fleet to the M113A3 standard. In U.S. Army terms, this modernization is known as the RISE program (Reliability Improvement of Selected Equipment). The M113A3 conversion was the result of a future system development program carried out by TARADCOM (Tank Automotive Research and Development Command) of Warren, Michigan. The prototype vehicle, which was known as the M113A1E1, was type-classified standard after intensive trials as M113A3 in 1987. Fielding of the first modified M113A3 by the U.S. Army took place in the same year. Major improvements of the M113A3 were the new turbocharged 5.2-liter 6V53T diesel engine that develops 275hp, a new Allison X-200-4A automatic transmission, hydrostatic steering, new driver controls using a steering wheel and brake pedal, and a new 200 amperes generating system with four 12-volt batteries. The external armored fuel tanks became a standard item (freeing 16 cubic feet of usable space inside the vehicle), a spall suppression liner was mounted, and all vehicles were prepared to be fitted with passive add-on armor. In addition, the M163 "Vulcan" trim vane became standard issue for all M113A3 series vehicles.

The conversion of the U.S. Army M113A1 and M113A2 to M113A3 was carried out between 1987 and 1994 by the Red River Army Depot in the USA, the Mainz Army Depot in Germany and the U.S. Army in Korea. Since 1994, conversions have been done by United Defense LP in the USA. The A3 RISE conversion was also carried out on the M113 family members such as the M577 and the M1059. All vehicles produced by United Defense LP since 1987 are of A3 standard. From the outside, the M113A3 can only be distinguished from the M113A2 by the provisions for the installation of the add-on armor kit.

A Tour around the M113A3 APC

Like its family predecessors, the M113A3 is a fully tracked, light, armored personnel carrier that provides protected transportation and cross-country mobility for troops and their equipment. Its box-shaped hull is made of aluminum armor, which provides the crew with protection against the effects of small arms fire and shell splinters. Inside the M113A3, a spall suppression lining is mounted to the side roof and rear. The power train is situated at the front right of the vehicle. It consists of the 5.2-liter, Detroit Diesel Engine Division of General Motors, water-cooled, turbo-charged, 6V53T two-stroke-cycle-type diesel engine that develops 275hp and the Allison X-200-4A Automatic hydrokinetic transmission with four forward and two reverse gears. The new transmission allowed the elimination of the transfer case and controlled differential used in earlier M113 variants.

The RISE power train increased the fuel economy, performance and braking capabilities of the vehicle. In addition, it allows the vehicle to maintain speed when driving on curves by accelerating the outer track instead of braking the inner track as in previous variants. All air intakes and outlets, including the exhaust, are placed under a grill assembly situated on the front right side of the roof. Located under the trim vane mounted on the front plate is an engine access hinge that opens to the upper front. The driver sits at the front left in the hull. He enters his position via a large one-piece hatch that opens to the rear. The M19 infrared periscope situated in the driver's hatch was replaced on the M113A3 with an AN/VVS-2 night viewer system. The M113A3 has a steering yoke that combines all steering controls in place of the steering laterals of earlier variants. The driver has a foot pedal for braking. Other controls situated at the driver's position include accelerator pedal, transmission control, ramp control lever, and warning light panel. The driver can also operate an engine compartment fire extinguishing system.

The position of the vehicle commander is situated in the center of the M113, behind the driver and power train. The roof-mounted commander's cupola, which can be traversed 360°, features five M17 periscopes, as well as a single-piece hatch that opens to the rear. Also belonging to the commander's cupola is a pintle mount for the 12.7mm M2 HB heavy machine gun. While the machine gun is the only armament of the basic carrier, most M113A3 variants are also fitted with a pair of smoke dischargers mounted on the left and right of the vehicle's front plate. Each side features a pack of four smoke discharger tubes.

The M113 family of vehicles has a running gear consisting of five road wheels, a rear idler and a front drive sprocket on either side. Torsion bar suspension supported by shock absorbers on the first, second and fifth road wheel is another feature of the M113A3. Earlier M113 models used shock absorbers only on the first and last road wheel on each side. The troop compartment of the M113A3 is situated in the rear of the vehicle, and there is space within for up to eleven soldiers and their kit. The soldiers enter and leave the vehicle via a large power-operated rear ramp. When the ramp is closed, the troops can also use a door in the ramp. Inside the troop compartment, the soldiers sit facing inwards on detachable benches located along the left and right side.

In the roof above the troop compartment there is a large one-piece hatch that opens to the rear. When opened, this allows the troops inside to fire over board of their vehicle, but this means the soldiers' upper bodies are exposed to enemy fire. The external armored fuel tanks are outside the hull to the left and right of the rear ramp. Each tank contains 180 liters (47.5 gallons) of diesel fuel. Like its earlier versions, the M113A3 is fully amphibious. When operated in water, the vehicle is driven and steered by its tracks, the same as it is on land. For amphibious operations, the trim vane is erected, which normally is folded back against the front plate. Two bilge pumps take care of ingressing water.

M113 in Foreign Service

Virtually since the time series production started in 1960, the M113 has been sold to other user nations. In the 1990s, the M113 and its variants were in service with the armed forces of the following nations: Argentina, Australia, Bahrain, Belgium, Bolivia, Bosnia and Herzegovina, Brazil, Cambodia, Canada, Chile, Colombia, Congo, Denmark, Ecuador, Egypt, Ethiopia, Germany, Greece, Guatemala, Iran, Iraq, Israel, Italy, Jordan, South Korea, Kuwait, Lebanon, Libya, Morocco, New Zealand, Norway, Pakistan, Peru, Philippines, Portugal, Saudi Arabia, Singapore, Somalia, Spain, Sudan, Switzerland, Taiwan, Thailand, Tunisia, Turkey, Uruguay, Vietnam, and Yemen. While the M113 has mostly been replaced in its role as armored personnel carrier, it is used by these nations in various roles. Most of these user nations have modified the originally delivered M113s to their own requirements. In addition, several users have built their own variants based on the M113 chassis and hull. In Germany, for example, Henschel *Wehrtechnik*, formerly known as Thyssen Henschel and today belonging to Krauss-Maffai Wegmann, developed an artillery observation post vehicle for the *Bundeswehr* based on the M113.

Not all M113s were produced at the San Jose facility in the USA by FMC, which today belongs to United Defense LP. Otobreda carried out license production in Italy and built several thousand M113s there. In addition, M113 kits were sent to foreign users and then assembled in country, example, in Pakistan. M113s were also produced in Turkey by Aiken, Steel Production Division, and FMC-Nurol. The huge number of user countries and variants make it nearly impossible to list the operational deployments of the M113 vehicle family. However, you can say that since the early 1960s, the M113 and its variants have taken part in every major conflict fought on this planet. The M113 was widely used by the U.S. and South Vietnamese forces during the Vietnam War, with Israeli forces in the 1967 Six-Day War, during the civil war in Lebanon in 1975, in the Yom Kippur War of 1973, and with U.S. and Coalition forces in the Gulf War of 1991. From 1992 onwards, vehicles of the M113 family have seen action with various nations in the former Yugoslavia. M113s have belonged to the inventory of the U.N. Forces operating in Bosnia and Croatia since 1992, of the IFOR and SFOR troops since 1995, and KFOR troops since 1999. The M113 also saw action in Somalia with U.N. forces in 1992, Rwanda in 1994 and East Timor in 2000.

Abbreviations

ACAV	Armored Cavalry Assault Vehicle
APC	Armored Personnel Carrier
APCLE	Armored Personnel Carrier Life Extension
ARVN	Army of the Republic of Vietnam
AMF(L)	ACE (Allied Command Europe) Mobile Force (Land)
CMTC	Combat Maneuver Training Center
DAREOD	Damaged Airfield Reconnaissance Explosive Ordnance Disposal
EFT	External Fuel Tank
EOD	Explosive Ordnance Disposal
FLIR	Forward Looking Infrared
FMC	Food Machinery Company
IFF	Identification Friend Foe
IFOR	Implementation Force
IFV	Infantry Fighting Vehicle
FISTB	Fire Support Team Bradley
FISTV	Fire Support Team Vehicle
ISAF	International Security Assistance Force
ITV	Improved TOW Vehicle
KFOR	Kosovo Force
MNB(S)	Multinational Brigade (South)
MNB(W)	Multinational Brigade (West)
MoD	Ministry of Defense
MTVL	Mobile Tactical Vehicle Light
NATO	North Atlantic Treaty Organization
REFORGER	Return of Forces to Germany
SAM	Surface-to-Air Missile
SFOR	Stabilization Force
SHORAD	Short Range Air Defense
SICPS	Standard Integrated Command Post System
TOW	Tube-launched, Optical-tracked, Wire-guided anti-tank missile
TARADCOM	Tank Automotive Research and Development Command
TUA	TOW Under Armor
UN	United Nations
UNPROFOR	United Nations Protection Force
USA	United States of America

Acknowledgements

First, I would like to thank my wife for her patience and support while I was working on this book. Many thanks also go to Michael Jerchel, Ralph Zwilling, Yves Debay, Peter Blume, and Walther Böhm for providing additional pictures. I would also like to thank JP Morgan and Clemens Niesner for additional support. I would also like to thank Freddie Leung for trusting me that I would be able to deal with the subject. I hope it is up to the required standard.

Technical Data of the M113 Basic Vehicle

Vehicle Variant	M113	M113A1	M113A2	M113A3
Series production:	since 1960	since 1964	since 1979	since 1987
Crew:	2 + 11	2 + 11	2 + 11	2 + 11
Combat weight:	10258kg	10750kg	11341kg	12247kg
Net weight:	8960kg	9702kg	9926kg	10832kg
Length:	4.86m	4.86m	4.86m	5.3m
Width:	2.69m	2.69m	2.69m	2.69m
Height to commander's cupola:	2.22m	2.22m	2.22m	2.22m
Maximum speed on land:	64km/h	67km/h	67km/h	66km/h
Maximum speed amphibious:	5.8km/h	5.8km/h	5.8km/h	5.8km/h
Track width:	0.38m	0.38m	0.38m	0.38m
Suspension type:	Torsion bar and two shock absorbers on each side at the first and fifth road wheel	Torsion bar and two shock absorbers on each side at the first and fifth road wheel	Torsion bar and three shock absorbers on each side at the first, second and fifth road wheel	Torsion bar and three shock absorbers on each side at the first, second and fifth road wheel
Engine:	5.9-liter Chrysler V-8 75M water-cooled petrol engine of four-stroke cycle type developing 209hp	5.2-liter Detroit Diesel Engine Division General Motors water-cooled diesel engine 6V 53 of two-stroke cycle type developing 215hp	5.2-liter Detroit Diesel Engine Division General Motors water-cooled diesel engine 6V 53 of two-stroke cycle type developing 215hp	5.2-liter Detroit Diesel Engine Division General Motors water-cooled turbo-charged diesel engine 6V53T of two-stroke cycle type developing 275hp
Transmission:	Allison TX-200 Automatic transmission with six forward gears and one reverse gear.	Allison TX-100-1 Automatic transmission with three forward gears and one reverse gear.	Allison TX-100-1 Automatic transmission with three forward gears and one reverse gear.	Allison X-200-4A Automatic hydrokinetic transmission with four forward gears and two reverse gears.
Armor of basic hull:	5083 Aluminum	5083 Aluminum	5083 Aluminum	5083 Aluminum
Cruising range:	321km	480km	880km	483km
Fuel capacity:	302 liters	360 liters	360 liters	360 liters
Slope:	60%	60%	60%	60%
Side slope:	30%	30%	30%	30%
Vertical obstacle crossing:	61cm	61cm	61cm	61cm
Trench crossing:	1.68m	1.68m	1.68m	1.68m
Ground clearance:	0.43m	0.43m	0.43m	0.43m
Batteries:	2x12 Volt	2x12 Volt	2x12 Volt	4x12 Volt
Alternator:	100 Amperes	100 Amperes	100 Amperes	200 Amperes
Acceleration 0-20mph:	12.0 seconds	10.5 seconds	11.7 seconds	8.1 seconds
Average cross country speed:	Data not available	Data not available	16.8 mph	22 mph

U.S. M113

This M113A2, which belongs to the 3rd Battalion, 12th Infantry Regiment (Mechanized) undergoes practice during an exercise held in southern Germany. Mounted on the roof of the vehicle next to the .50-caliber M2HB machine gun is an M47 Dragon anti-tank guided weapon. The shield in front of the weapon does not provide any armor protection but prevents the weapon from being damaged by branches when operating in wooded area. Note the MERD- type camouflage pattern missing one color. (Peter Blume)

Close-up of the M47 Dragon anti-tank guided weapon mounted next to the .50-caliber M2HB machine gun on an M113A2 APC of the U.S. Army. Mounting the M47 Dragon on the roof of the M113A2 APC was a common practice in order to allow the crew to engage enemy armor without dismounting. The M47 Dragon has a combat range of 65 to 1000m (71 to 1093 yards) and is able to penetrate up to 60cm (24 inches) of conventional steel armor.

As we have given most of the development history and modification of the U.S. M113 in the introduction of this book, we will keep it short here. By the beginning of the 1990s, the M2 Bradley Infantry Fighting Vehicle (IFV) had replaced the M113 in service with the U.S. Army in its most important role: vehicle for the mechanized infantry. However, in the U.S. National Guard the M113 remained the battle taxi for several infantry units. Like the armed forces of other nations, the U.S. Army also uses the M113 in countless variants as a utility tracked armored combat vehicle. During the period covered by this book, the entire U.S. Army fleet of M113s was updated to the M113A3 standard. With the A3 RISE modification there was also a change of numbers among some of the M113 vehicle fleet vehicles. The M106A2, which was fitted with a 120mm mortar instead of the 107mm, is now called the M1064A3. Upgraded with a special set of command and control equipment, the M577A2 is now called the M1068A3 Standard Integrated Command Post.

In the 1990s, the U.S. armored mechanized infantry and artillery units used the M113 mostly in its command post role, as an armored ambulance and as a mortar carrier. It still serves as an engineer section vehicle in the engineer units. In addition, a lot of special M113 variants were in service and still are, but some like the M730 Chaparral or the M163A1 Vulcan were replaced by more advanced systems, for example the HMMWV-based Avenger air defense system. In the late nineties it was estimated that the U.S. Army still operated a fleet of 25,000 M113 and M113-based vehicles.

In the foreground of this photo is an M113A2; an M901A1 Improved TOW Vehicle is visible in the background. Both vehicles belong to the 3rd Battalion, 12th Infantry Regiment (Mechanized) and were seen in 1990. Belonging to the battalion HHC (Headquarters and Headquarters Company), the M113A2 in front is fitted with the Armored Cavalry Armored Vehicle kit. This features an armor kit and shield for the commander's cupola, as well as two shields and two M-60 machine gun mounts for the rear compartment. (Peter Blume)

The M901 ITV (Improved TOW Vehicle) entered service in 1978, and more than 3100 were produced for the U.S. Army. Basically, the vehicle consists of the M113A1 chassis and the M27 cupola with the TOW launcher, which can be elevated. Mounted inside the M27 cupola are two TOW launcher tubes, the TOW sight assembly and the target acquisition sight. The M27 cupola can be traversed a full 360°, while the launcher part has an elevation of +34° to –30°. The U.S. Army phased the vehicle out of its inventory in the 1990s. The vehicle's TOW launcher is lowered to the travelling position.

Seen during the REFORGER exercise "Centurion Shield 90," this M577A2 command post vehicle displays the MERDC camouflage pattern used by U.S. forces in central Europe between the mid-1970s and the introduction of the standard three-color NATO camouflage pattern in the mid-1980s. Re-painting in the U.S. Army units took time, and often vehicles from the depot that bore the MERDC camouflage scheme remained so for several years. Even during the 1996 IFOR operation and the 1999 KFOR deployment there were still U.S. vehicles and M113 variants such as the M577 that were painted in MERDC pattern.

This M577A2 command post vehicle has its trim vane lowered and its engine access hatch opened for maintenance work on the vehicle's engine. The vehicle belongs to the 5th Battalion, 17th Field Artillery Regiment of the 210th Field Artillery Brigade. In 1990, when the picture was taken, the unit belonged to the corps artillery of the VIIth Corps and was equipped with the M110 self-propelled howitzer. In artillery units, the M577A2 operates as a command post, but it also sees use as a fire-direction center vehicle.

This picture shows an M1015 carrier for containerized equipment, which is based on the M548A1. In total, 252 vehicles saw service with the U.S. Army between 1982 and 1996. This vehicle of the 522nd Military Intelligence Battalion of the 2nd Armored Division, which is taking part in the REFORGER 90 exercise "Centurion Shield," carries an AN/TSQ 114 Trailblazer radio communications intercept and direction finder system.

The pictured M163A1 Vulcan belongs to the 5th Battalion, 3rd Air Defense Artillery Regiment of the 8th Infantry Division. The vehicle chassis, which is designated as M741, is basically an M113A1 fitted with a suspension locking system, buoyant pods on either side of the hull, a buoyant trim vane, and an additional hatch on the vehicle's roof. Mounted in the center of the roof is the the power-operated one-man turret with the 20mm M168 Vulcan gun. (Michael Jerchel)

For several years, the M163A1 Vulcan formed the close air defense capability of the U.S. Army. Together with the M730 Chaparral, it was operated by the Air Defense Artillery Regiments of the combat divisions. These battalions were equipped with 24 pieces of each system. Other nations' military that use the M163A1 are the armed forces of Israel, Portugal, Jordan, Morocco, Thailand, Sudan, and Yemen. (Michael Jerchel)

The main weapon system of the M163A1 Vulcan is the 20mm M168 Vulcan Gun. The gun can be fired either with a cadence of 1000 rounds per minute when used against ground forces or with 3000 rounds a minute when used in the air defense role. Maximum combat range against flying targets is 1200m (1312 yards). Under the left cover next to the gun is situated the EMTECH AN/VPS-2 I-band pulse Doppler radar. In addition to the 20mm gun system inside the vehicle, two Stinger SAMs and a shoulder launcher are also carried. (Michael Jerchel)

Seen in 1993, this M577A2 belongs to the medical platoon of 3rd Squadron, 4th Cavalry Regiment of the 3rd Infantry Division. (Ralph Zwilling)

In addition to being used as a command post and fire direction center, the M577A2 also saw service as a medical treatment facility in armored units. This vehicle belongs to the 1st Battalion, 63rd Armor Regiment of the 1st Infantry Division. (Ralph Zwilling)

The crew of this M577A2 raises the tent extension in order to provide improved space for a medical treatment facility. (Ralph Zwilling)

The crew of the M1059A3 Carrier Smoke Generator consists of the driver, vehicle commander and smoke generator operator. The M157A2 smoke generator set consists of two M54A1E1 smoke generator assemblies, the 438-liter (116-gallon) fog oil tank mounted inside the vehicle behind the commander's station, an air compressor, the fog oil pump, the control panel assembly, and two 50-gallon fuel (189-liter) cans mounted outside on the EFTs. The M54A1E1 smoke generator is a pulse jet engine that vaporizes fog oil in order to provide a thick smoke screen.

In 1996, the U.S. Army ordered 288 M157A2 smoke generator systems and 342 Modification Kits to upgrade the Army's inventory of M1059 smoke generator carriers. The M1059 Carrier Smoke Generator consists of an M113A2 armored personnel carrier onto which the M157 smoke generator system is mounted. The modified M157A2 remote-controlled unit uses a pulse jet engine to produce the smoke. In addition to the new M157A2 generators, the chassis of the vehicle was brought to the M113A3 standard, and the new vehicles were called M1059A3s. The M157 Smoke Generator Set has been in service with the U.S. Army since 1986 and the M1059 since 1988.

The M1059A3 Carrier Smoke Generator seen here was deployed to Tuzlar airfield in Bosnia in early 1996. The vehicle belongs to the 25th Chemical Company of the 1st Armored Division then deployed under IFOR in Bosnia. The U.S. Army purchased a total of 323 M157 Smoke Generator Sets (SGS) and 276 M1059 Carrier Smoke Generators.

Even though the M2 Bradley replaced the M113 in its classical role as armored personnel carrier for the infantry, in the nineties the M113 still was present in all armored units of the U.S. Army, performing a wide range of tasks. These M113A3 engineer section vehicles of the 23rd Engineer Battalion were seen during the Bosnia deployment of the unit with IFOR in early 1996. The IFF panels mounted to the sides, which first saw use in Bosnia, are a result of several friendly fire incidents that occurred during the 1991 Gulf War. Unfortunately, the system is not as effective as had been hoped, as proven in the 2003 Gulf War, where U.S. and British forces once again suffered a number of casualties from friendly fire.

This M113A3 armored ambulance belongs to 3rd Squadron, 5th Cavalry Regiment of the 1st Armored Division and was seen during the deployment of the unit to Bosnia in 1996 as part of IFOR. Note the buoyant trim vane, which became necessary with all M113A3 variants in order to tackle the changed point of gravity of the vehicle after the addition of the EFTs (External Fuel Tanks).

The M548A1 is a tracked cargo carrier based on the M113A1. The first M548 entered the U.S. Army's inventory in 1966. From 1984 onwards, the vehicles were modernized to the M548A1 standard, and a further upgrade program was carried out beginning in 1993, bringing the vehicle to the M548A3 standard. The different improvements usually were similar to those carried out on the U.S. Army's M113s. The photograph shows an M548A1 of 23rd Engineer Battalion of the 1st Armored Division bogged down during the Bosnia deployment of the unit as part of IFOR in 1996.

An M548A1 is shown during the recovery of a second vehicle somewhere in Bosnia in March 1996. Both vehicles belong to 23rd Engineer Battalion of 1st Armored Division. Based on the M113A1 chassis, the M548A1 is a tracked cargo carrier. The vehicle has a crew of four soldiers and a combat weight of 1.8 tons. The payload of the vehicle is 5 tons, and the cargo bay has a dimension of 3.3m (10.8 ft) by 2.4m (7.8 ft). The fully amphibious vehicle shares many technical details with the M113, including engine, gearbox and running gear.

This M577A2 command post vehicle, which belongs to the Headquarters and Headquarters Company of the 1st Armored Brigade of the 1st Armored Division, was seen during the IFOR deployment of the unit in Bosnia in early 1996. An IFF panel is on the side of the vehicle, and above the generator is an additional storage box that the crew placed there.

In the U.S. Army, the forward observation officer teams of the artillery use the M981A3 FISTV. The vehicle entered service in early 1985. Since 1997, when the production of the M7 Fire Support Team Bradley (FISTB) started, the days have been numbered for the M981A3 FISTV's, but it is estimated that the vehicle will stay in service well into the first decade of the 21st century. The pictured vehicle belongs to the division artillery of the 1st Armored Division and is attached to the 1st Squadron, 1st Cavalry Regiment.

An M577A2 during the 1996 IFOR deployment of U.S. forces to Bosnia. Note the IFF panels and the raised antenna mast.

Deployed in the defense perimeter of a U.S. IFOR camp in Bosnia, an M981A3 FISTV of 2nd Battalion, 3rd Field Artillery Regiment can be seen in 1996. The vehicle's raised turret contains the AN TVQ ground/vehicle laser locator designator (GVLLD).

An M981A3 FISTV (Fire Support Team Vehicle) of 1st Battalion, 7th Field Artillery Regiment. This rear view shows to advantage the EFTs that were mounted to all M113 of the U.S. Army under the A3 RISE upgrade program. The vehicle's turret is lowered into the travelling position.

Rear view of an M1059 Carrier Smoke Generator of 25th Chemical Battalion belonging to the 1st Armored Division deployed to Bosnia as part of IFOR in 1996. Note the two M157 Smoke Generator Sets, which can produce a fog screen over a period of an hour using the fog oil stored in the vehicle. The vehicle is fitted with an ACAV commander's up-armored cupola, and its front shield can be seen on the rear roof of the vehicle.

This M113A3, which belongs to the 9th Engineer Battalion of the 1st Infantry Division, took part in the KFOR deployment of the unit in June 1999. The mounting points for the add-on side armor are clearly visible. Used as an engineer section vehicle, this M113A3 has a 40mm Mk.19 automatic grenade launcher mounted in front of the commander's cupola. Since the Mk.19 entered service with U.S. forces, the weapon has been a common sight on vehicles. Usually a mix of Mk.19s and .50-caliber M2HB machine guns is mounted to the vehicles of a unit.

An M577A2 belonging to 1st Battalion, 6th Field Artillery Regiment undergoes engine maintenance. It looks as if the mechanics will remove the engine completely using the M88A1 armored recovery vehicle. The photo was taken in Kosovo in July 1999. At that time, both vehicles belonged to the U.S. KFOR contingent.

Seen during Exercise "Rolling Steel 99," this M577A2 has its tent extension connected. The tent connection provides further working space for HQ staff and it is possible to connect four vehicles in a cross formation to form one big HQ complex. The vehicle belongs to the 2nd Battalion, 3rd Field Artillery Regiment of the 1st Armored Division. Note the additional equipment stored on the vehicle and the roof-mounted antennas.

This overhead view of an M577A2 of the 1st Battalion, 77th Armor Regiment of the 1st Infantry Division allows us to see the commander's hatch, antenna mounts and the stored manual light crane equipment that can be used to replace the front-mounted 28-volt generator when it is required for dismounted use. Note that only three of the available six antenna sockets are used.

Mechanics have removed the engine of this M577A2. The cooling assembly and the 5.2-liter Detroit Diesel Engine Division General Motors water-cooled 6V 53 two-stroke cycle-type diesel engine can be seen placed in front of the vehicle. The troops belong to the U.S. KFOR contingent. This picture was taken in Kosovo in July 1999.

An M981A3 FISTV (Fire Support Team Vehicle) crosses a ribbon bridge during Exercise "Dangerous Crossing 01" in southern Germany. The vehicle belongs to the 1st Battalion, 7th Field Artillery Regiment, of which the main weapon system is the M109A6 Paladin self-propelled howitzer. The vehicle is painted in the standard NATO three-color camouflage.

The U.S. Army's engineer units use the M113A3 as a section vehicle. This vehicle belongs to the 9th Engineer Battalion of the 1st Infantry Division. It is towing an M200 trailer fitted with an M58A1 Mineclearing Line Charge during a CMTC rotation in 2001. Each engineer company of the battalion has nine M113A3s that used as engineer section vehicles. The pictured vehicle is fitted with the MILES II duel simulation system, and the crew's equipment can be seen stored on the outside. (Ralph Zwilling)

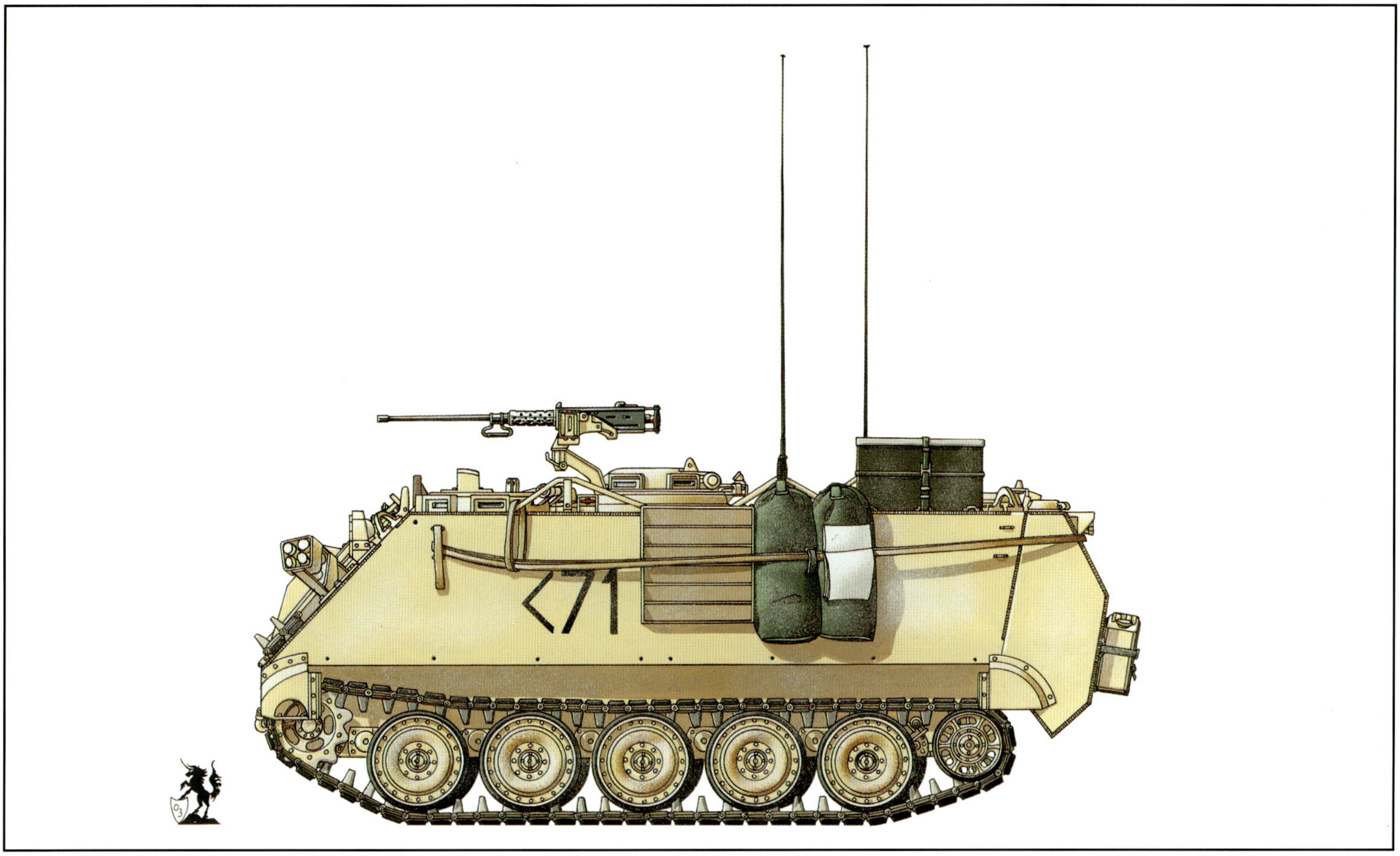

When U.S. troops began the invasion of Iraq in March 2003, they still had a large number of M113s in their inventory. This drawing shows an M113A3 of the HHC company of a unknown mechanized infantry battalion of the 3rd Infantry Division on its way to the Iraqi border. Note the IFF panel on the side of the vehicle side and the way the crew's equipment is stored.

Rear view of an M113A3 of the 1st Infantry Division's 9th Engineer Battalion during its CMTC rotation in Hohenfels 2001. The vehicle is towing an M200 trailer fitted with an M58A1 Mineclearing Line Charge. The system clears a minefield by firing a line charge containing more than 800kg (1764 lb.) of explosives over the minefield and detonating it when in place. This can clear a path 12m (39 feet) wide and 100m (328 feet) long. (Ralph Zwilling)

The U.S. Army deploys the M139 Volcano Vehicle-Launched Scatterable Mine System (VLSMS), which is based on an M548A1. The M139 Volcano multiple delivery mine system makes it possible to lay 960 mines at a density of 0.9mines/m. Either anti-tank or anti-personnel mines can be used, and a typical M87 Minecanister mix includes one BLU-92B anti-personnel mine and five BLU-91B anti-tank mines. (Ralph Zwilling)

The M548A1 fitted with the M139 Volcano Vehicle-Launched Scatterable Mine System (VLSMS) belongs to the 9th Engineer Battalion of the 1st Infantry Division and was photographed during the unit's CMTC rotation in 2001. (Ralph Zwilling)

Rear view of the same M548A1 fitted with the M139 Volcano Vehicle-Launched Scatterable Mine System (VLSMS) belonging to 9th Engineer Battalion. The U.S. Army operates about 300 M139 Volcano VLSMS based on the M548A1. Additional systems are mounted on 5-ton trucks, in helicopters and on trailers. (Ralph Zwilling)

This M113A3 armored ambulance belongs to 1st Squadron, 1st Cavalry Regiment, which is the armored reconnaissance asset of the 1st Armored Division. The M113s used in the ambulance role are fitted with a stretcher kit that allows for carrying up to four casualties on stretchers. Note the IFF panels mounted on the vehicle's side.

This photograph of an M113A3 ambulance of the 1st Battalion, 77th Armor Regiment provides a good view of the running gear. The running gear of the M113 family of vehicles consists of five road wheels, a rear idler and a front drive sprocket on either side. The M113 has torsion bar suspension supported by shock absorbers on the first, second and fifth road wheel.

Another picture of an M113A3 armored ambulance. This vehicle belongs to the 2nd Battalion, 6th Infantry Regiment, which is part of the 1st Armored Division. Photographed during an exercise in the autumn of 2001, the vehicle shows the mounting points for the bolts of the add-on side armor, which made up part of the RISE program modifications. Note the layout of the roof of the M113A3; the driver's hatch is situated to the front left, while the commander's hatch is situated in the center. The engine air intake louvers and exhaust are visible at the front right.

An M577A2 belonging to the 1st Squadron, 1st Cavalry Regiment on exercise in central Germany in 2002. The vehicle has one of its antenna masts mounted on the roof, while other antennas are close to it on the raised part of the roof.

An M577A2, which belongs to the 1st Squadron, 1st Cavalry Regiment, takes part in a field training exercise in 2002. The vehicle features the standard NATO three-color camouflage pattern. The history of the M577 began with the building of the first prototypes in 1962. These were simple conversions of M113s with their hulls raised over the troop compartment. Series production started the same year. More than 7000 vehicles were manufactured in the basic, A1, A2 and A3 versions, and they are in service with forces all over the world. From 1992 onwards, the U.S. Army modified their fleet of M577s to the M1068A3 Carrier Standard Integrated Command Post System (SICPS).

An M1064A3 mortar carrier of 1st Squadron, 1st Cavalry Regiment takes a break during a field training exercise in the summer of 2002. In the cavalry units, two M1064A3s are located with each of the three ground troops of the squadron. In the armored and mechanized infantry units, the M1064A3 is located in the mortar platoon of the HHC. Note the EFTs and the open door in the rear ramp.

Soldiers belonging to the mortar platoon of 1st Battalion, 77th Armor Regiment use a 120mm mortar during a fire mission. Previously known as the M106, the vehicle was designated as M1064A3 following the introduction of the 120mm mortar.

There are more than 1000 M1064A3 mortar carriers in service with the U.S. Army. Each armored battalion, mechanized infantry battalion and cavalry squadron is equipped with six M1064A3s. Note the base plate on the right vehicle side, which allows for using the mortar outside the vehicle in a dismounted role. The pictured M1064A3 belongs to 1st Battalion, 77th Armor Regiment. The M1064A3 has a crew of four consisting of the driver, commander, gunner, and loader. As many as 69 120mm mortar bombs can be stored within the vehicle.

An M981A3 FISTV supports the 1st Squadron, 1st Cavalry Regiment of the 1st Armored Division during an exercise in 2002. The M981A3 is the upgraded version of the M981A1 FISTV, which in turn is quite similar to the M901A1 ITV tank killer. However, instead of being fitted with the TOW missile, the turret of the vehicle, which is in the raised position, is equipped with sophisticated observation and surveillance equipment such as the AN/TAS-4 thermal sight and the AN TVQ G/VLLD sighting system. Also fitted to the M981 FISTV is an extended radio pack, as well as a vehicle navigation system.

Canadian M113

A pair of M113A1 APCs belonging to the 1er Bataillon Royal, 22me Régiment *is seen during an exercise in Germany. The vehicles are fitted with the MILES duel simulation equipment. Before wheeled APCs such as the LAV entered the Canadian Armed Forces' inventory, the M113 had been the most important vehicle of the mechanized infantry units. (Peter Blume)*

In the 1990s, Canada operated a fleet of approximately 1200 vehicles of the M113 family. The M113 and its variants entered service with the Canadian Armed Forces in 1965. In 1968, the M114 Command and Reconnaissance Carrier, which is based on the M113, was also introduced to the forces' inventory; in Canada the M114 C&R is called Lynx. In 1985, the original Canadian M113 fleet was bolstered by the purchase of M113A2s and the remaining M113A1s were upgraded to the M113A2 standard. Most of the Canadian M113A2s were fitted with EFTs (External Fuel Tank). The Canadian M113 fleet includes APCs, the M577A2 command post vehicle, M579A2 fitters vehicle, the M548A1 tracked cargo carrier, and the M113A2 engineer vehicle. In addition to these commonly known M113 variants, the Canadian Army operates two unique versions. Exclusively used by Canada, these are the M113A2 DAREOD (Damaged Airfield Reconnaissance Explosive Ordnance Disposal) vehicle and the M113A2 ADATS air defense missile system.

Like Norway, Canada also operates the M113A1 fitted with a one-man Kvaerner Eureka turret armed with the TOW anti-tank weapon system. This vehicle is known as the M113A1 TUA (TOW Under Armor) and a total number of 72 exist in the Canadian Armed Forces' inventory. Like the other M113 family variants, the M113A1 TUA were also upgraded to A2 standard. At the time of writing, the Canadian M113 Armored Personnel Carrier Life Extension Program is underway. The M113APCLE includes the conversion of a fleet of 341 M113A2s to M113A3 standard while 61 additional vehicles are short-listed for the same conversion. A total of 183 of the 341 vehicles will be converted to the stretched M113 version, the MTVL (Mobile Tactical Vehicle Light). Then the Canadians will use the MTVL in a number of variants, the way that the M113A2 was before.

In the late nineties, the LAV 25 Coyote replaced the M114 Command and Reconnaissance Carrier in active Canadian Army service. Currently, the replaced vehicles are in reserve. In total, Canada received 174 Lynx, the first of which was delivered in 1968. This M114 C&R, which belonged to the Canadian forces in Germany, was seen during Exercise "Royal Sword 90" in the early 1990s shortly before the Germany-based Canadian forces re-deployed to Canada. (Peter Blume)

The M114 Command and Reconnaissance Carrier is based on components of the M113A1. However, the engine of the vehicle was transferred to the rear, its hull is much lower than that of the M113A1 and there are only four road wheels on each side. The Lynx has a crew of three, a combat weight of 8.7 tons and a maximum road speed of 68km/h (42 mph). Like the M113A1, the Lynx is amphibious. The vehicle has a fuel capacity of 303 liters (80 gallons) and a road range of 523km (325 miles). The Canadian Lynx is armed with a 12.7mm and a 7.62mm machine gun. (Yves Debay)

Like most users of the M113 vehicle family, the Canadian army also uses the M577A1 command post vehicle. Pictured in the early '90s during a field training exercise, this M577A1 belonged to the Canadian forces based in Germany. Worthy of note are the roof-mounted additional storage box with fastenings for an antenna mast and the wire cutters mounted to the front of the hull. (Peter Siebert)

The Canadian Armed Forces operate several M579A1 fitters vehicles together with their M113 fleet. The M579A1 is fitted with a HIAB 1.5-ton hydraulic crane with a reach of 3 meters (10 feet). The vehicle's roof is modified so that the commander's copula and the hatch over the troop compartment are mounted on a huge roof hatch that can be opened to the right. This makes it is possible to carry a second M113 engine pack inside the M579A1. An additional floating panel on the trim vane compensates the weight of the hydraulic crane during use of the vehicle in its amphibious role. (Picture: Peter Siebert)

From 1988 onwards, the Canadian Armed Forces received 36 M113A2 ADATS air-defense missile systems as part of a new low-level air defense network (LLAD). Manufactured by Oerlikon–Contraves and based on an M113A2, the ADATS system consists of the traversable remote-controlled turret with FLIR, SHORAR radar, TV tracker, laser range finder, missile guidance laser, and four missile containers mounted on each side of the turret. The ADATS missile can destroy targets up to an elevation of 6km (3.7 miles) and a distance of 10km (6.2 miles). (Walter Böhm)

The ADATS operators are seated in the back of the M113A2 chassis in front of two operating consoles. Using the rotating SHORAR radar, the crew is able to detect enemy aircraft up to a distance of 24km (15 miles). The M113A2 chassis used to mount ADATS were also fitted with EFTs. The M113A2 was extensively modified to accept the ADATS. (Peter Siebert)

During Exercise "Royal Sword 90," which was held in Germany, a Canadian Army M113A2 TUA takes up its new firing position. In the Canadian Armed Forces' inventory, the M113A2 TUA replaced the M113A1 with a TOW launcher mounted in the open large rear hatch. The TOW missile can destroy every known armor threat up to a distance of 3750m (4098 yards). Note the NATO three-color camouflage pattern on the vehicle. (Peter Blume)

Used by Canadian forces deployed to an AMF(L) exercise in Norway in the early nineties, this M113A1 is fitted with a hydraulic dozer blade. The armament is the common pintle-mounted .50-caliber M2HB heavy machine gun. Note that the vehicle is fitted with a transparent windshield in front of the driver's position in order to protect the driver from the freezing temperatures. (Yves Debay)

This M113A2 Recovery Vehicle belonged to the UNPROFOR troops in Bosnia in 1992. The M113A2 Recovery Vehicle features a 9-ton hydraulic winch and a spade system mounted at the rear of the vehicle. The M113A2 RV has a combat weight of 11.6 tons and a crew of three consisting of the driver, commander and winch operator. A manually operated hydraulic crane is situated on the left side of the vehicle's roof. The collapsible crane, which can be easily identified in the picture, has a lift capacity of 1.3 tons and a reach of 1.5 meters (5 feet).

The Canadian engineer units use the M113A2 Engineering Variant Specially Equipped Vehicle. Originally designed based on the M113A1, the Canadian vehicle was upgraded to the M113A2 standard. The M113A2 Engineering Variant Specially Equipped Vehicle features a hydraulically operated bulldozer blade, a hydraulically operated earth auger, a hydraulically operated tool system, and a ramp retaining device. The use of EFTs allows the storage of additional engineer tools inside the vehicle. This M113A2 Engineering Variant Specially Equipped Vehicle belonged to the Canadian UNPROFOR troops and was photographed in Bosnia in 1993.

The M113A2 Engineering Variant Specially Equipped Vehicle is fitted with a hydraulically operated earth auger, which is shown here in its traveling configuration on the left side of the roof. Operated by a two-man team, the hydraulic earth auger can drill holes with a diameter of 20cm (8 inches) and up to a depth of three meters (10 feet). Other hydraulic tools carried by the vehicle are a chain saw, jackhammer and impact wrench.

A Canadian UN patrol consisting of several M113A2s fitted with ACAV kits moves down a road in Bosnia in early 1993. Most Canadian M113A2s were fitted with External Fuel Tanks (EFT).

A Canadian UNPROFOR M113A2 fitted with the ACAV (Armored Cavalry Assault Vehicle) kit minus the commander's machine gun shield patrols in Kiseljak in early 1993. With the ACAV kits, the Canadian M113A2s look very similar to the vehicles of the U.S. Army used during the Vietnam War. Note that the recently fitted shields and armored weapon station are still in green camouflage paint, while the M113A2s are regulation UN white-with-black UN markings.

Taken through the open rear ramp of an M113A2 of the Canadian Armed Forces in Bosnia, this photograph illustrates well the large space available in the troop compartment of the vehicle. The driver's compartment is visible at the rear to the left, and the folded commander's seat is in the center. The crew benches are located to the left and right, and the crew's equipment is stored in the space above the tracks. The vehicle radio is mounted on the left behind the driver's position. The engine compartment is situated to the front right where the large warning panel is visible.

Close-up of a machinegun shield of a Canadian M113A2 ACAV. Shields like the one pictured were mounted on the vehicle's roof to the left and right of the large hatch above the troop compartment. The shields are identical to those used by the U.S. Army on their M113A1 ACAVs in Vietnam during the late 1960s.

In 1992, soon after they arrived in Bosnia, the Canadian UNPROFOR troops fitted their M113A2 APCs with ACAV (Armored Cavalry Assault Vehicle) kits. The kits consisted of a shield and armored weapon station for the commander and two machinegun shields. The design of the rear of the armored weapon station mounted to the commander cupola allows for opening the commander's hatch without limitations; if the hatch is open, the hatch protects the commander's back.

Early 1996: This M113A2 DAREOD belonged to the Canadian IFOR contingent based in Kljuc in Bosnia. Clearly visible is the 360°-turnable periscope mounted in the right side panel of the vehicle at the rear. A similar periscope is also located on the left side at the level of the driver's compartment. The DAREOD is fitted with a one-man turret in which sophisticated night vision equipment is located.

The M113A2 DAREOD can be fitted with additional searchlights on the front of the hull and on the sides of the vehicle. Situated on the left side of the hull at the driver's compartment is a 360°-turnable periscope that allows the driver to look down onto the surface next to the vehicle's track. Note that the vehicle is fitted with passive add-on armor.

Taken in March 1998, this photo shows an M113A2 DAREOD and an M113A2 Engineering Variant Specially Equipped Vehicle parked outside the Canadian forces' base near Kljuc. The M113A2 DAREOD is fitted with a Pearson Surface Mine Plough (SMP 1). The SMP 1 is a hydraulically operated, V-shaped, mine-plough system that physically removes mines from the surface in front of the vehicle. It was for use on airfield runways along with the M113A2 DAREOD.

The M113A2 TUA first saw service in Bosnia with the Canadian UNPROFOR troops in 1992. In 1996, M113A2 TUAs were still in service in the former Yugoslavia, this time with the Canadian IFOR troops. These M113A2 TUAs were seen in March 1996 outside the Canadian Forces Base near Kljuc.

The inside view of an M113A2 TUA. The stowage rack for the TOW missiles can be seen on the right side. The rack is fully stowed with seven missiles. Note that two additional missiles are carried in their wooden containers. The missiles used are TOW 2A BGM 71E-1B. Earlier versions of the vehicle, e.g., the M113A1 TUA, had a different layout of the interior and could carry fewer missiles. Inside the vehicle, the loader is seated on the bench on the left side, the gunner is positioned in the TOW turret and the commander is seated to the right of the turret.

Like the Norwegian Army, the Canadian Armed Forces use a tank killer based on the M113. The M113A2 TUA (TOW Under Armor) uses the same turret as the NM 142 Rakettpanserjager. The turret was developed by Kvaerner Eureka A/S and built in Canada under license. A total of 72 M113A1s were fitted with the TUA turret and issued to the Canadian Army. The vehicles were brought to A2 standard in the late eighties. The pictured vehicle, which belonged to the Canadian KFOR contingent, was seen in March 2000 in Kosovo.

A Canadian Army M113A2 TUA participating in Exercise "Caravan Guard 89," which was held in Germany, can be seen fitted with special exercise markings consisting of an orange circle with a black edge and the number "23". The vehicle belongs to the *1er Bataillon Royal, 22eme Régiment* of the 4th (CA) Mechanized Brigade. Within the mechanized infantry battalions, the M113A2 TUAs were organized into the fire support company, which could field 18 such vehicles. The vehicle is painted in the standard NATO three-color camouflage pattern.

Kosovo, March 2000: An M113A2 Engineering Variant Specially Equipped Vehicle of the Canadian KFOR contingent patrols near the Pristina airport. The vehicle is fitted with passive add-on armor, and a 7.62mm FN MAG machine gun is mounted on the pintle mount in front of the commander's cupola.

German M113

Between 1969 and 1970, the Bundeswehr *fitted 29 M113GA1s with the EMI Electronics mortar location radar No 8 Mk 2 Green Archer. The vehicle's hull was modified so the radar could be mounted in the rear of the vehicle. The radar, which has a range of up to 15km (9.3 miles), can measure the flight of the projectile and thereby detect the position of enemy mortars and artillery units. The 12.5-ton vehicle has a crew of four soldiers and measures 5.04m (16 ft) in length, 2.69m (8.8 ft) in width and 2.85m (9.4 ft) in height. The M113 with Green Archer radar has also been in service with the Danish Army and the Italian Army. The pictured vehicle, which belongs to* Beobachtungsbattalion *23, was seen on the Bergen Hohne ranges in 1989. By the mid-nineties, all M113GA1 Green Archer had been withdrawn from service. (Michael Jerchel)*

This Flieger-Leit-Panzer *(FlgLtPz) M113GA1 forward air controller vehicle is seen during the REFORGER exercise "Centurion Shield 90." The vehicle is fitted with a special VHF antenna and radio set to allow communication with fighter aircraft. The radio set consists of a VRC 240G, a GRC 106 G4 and an SEM 25/35. The FlgLtPz M 113GA1 was used in brigade and division level headquarters units in order to co-ordinate the air support for ground troops.*

Germany operates one of the largest fleets of M113 family vehicles outside the United States and introduced the M113 into its inventory as early as in 1962. The first batch of M113s comprised 1646 vehicles and was issued to armored infantry and engineer units. In the following years, the German MoD purchased more M113s, and by 1991 the German Army's inventory included no fewer than 3763 M113s in countless variants. Shortly afterward in 1993, the M113 was replaced completely in its role as APC with the Marder IFV, but the more than 3000 specialized vehicles remained in service.

In 1963, the *Bundeswehr* (German Armed forces) started to modify the delivered M113s to its requirements for the first time. The modified vehicles were designated M113G. From 1974 to 1981, all M113Gs were brought to the M113A1 standard and consequently called the M113GA1. The A1 upgrade program was followed with the A2 upgrade program in the 1980s. The *Bundeswehr* upgraded only a limited number of vehicles to the A2 standard and even fewer were fitted with EFTs. In addition to the standard typology (A1, A2, EFT, etc.) in the German Army, variants of the M113 are further described as follows:

- G or GE (both terms describe the same): Vehicle with German light system, a Diehl 213B track, new fire extinguishing system, Webasto heating system, smoke discharging system, German SEM 25/30 radios, MG 3 machine gun mount, and modified periscopes.
- A0: Vehicles fitted with the SEM 80/90 radio assembly.

The Lance ballistic missile system consisting of the Lenkraketenwerfer Startfahrzeug *(LRakWfSF) Lance M667 ballistic missile launcher is seen here in the background, and the* Lade-Transport-Panzer *(LadeTrspPz) Lance M668 missile loader transporter is in the foreground. Between 1976 and 1992, the* Bundeswehr *operated 26 vehicles of each kind. The Lance ballistic missile system was used in three rocket artillery battalions, each attached to one of the three corps. The units were* Raketenartilleriebataillon *150 in Wesel,* Raketenartilleriebataillon *350 in Montabaur and* Raketenartilleriebataillon *650 in Flensburg. The vehicle in the photo belongs to* Raketenartilleriebataillon *350. (Peter Blume)*

A Lenkraketenwerfer Startfahrzeug *(LRakWfSF) Lance M667 ballistic missile launcher is seen in the launch position with missile raised. The M782 launcher assembly, which is mounted inside the vehicle, can also be used on an M740 wheeled trailer. The ballistic missile launcher has a combat weight of 10.7 tons and is operated by a six-man crew. Both the M667 and the M668 are 6.58m (21.5 feet) long, 2.69m (8.8 feet) wide and 2.74m (9 feet) high when measured with the extended driver's compartment. Most of the other technical data is identical to that of the M548A1. (Michael Jerchel)*

Due to the heavy weight that the vehicles had reached by then, in the late '80s the German Army issued orders that the M113 family of vehicles was no longer to operate in the amphibious role. Subsequently, the trim vane was replaced on the M113s with a special storage cage for additional equipment. In the late 1980s, it became obvious that the M113 would remain in the service of the German Army until well after the year 2010.

But the earlier upgrades resulted in a heavier combat weight, which dramatically limited the performance of the M113. This made a further life extension program necessary. The program, originally called "*Nutzungsdauerverlängerung* (NDV)," was split into two parts: NDV 1 and NDV 2. The vehicles nearing the end of their service life received only NDV 1 to reduce costs. Both NDVs were carried out by the *Flensburger Farzeugbau Gesellschaft mbH*. The NDV 1 modification mainly consisted of the installation of a new two-circuit braking system including a mechanical hand break, a new steering dashboard and a new driver dashboard. NDV 1 was applied to all selected variants of the M113 family of vehicles with a combat weight less than 12.5 tons. NDV 2 was applied to vehicles with a combat weight of more than 12.5 tons. The NDV 2 included all modifications of the NDV 1 plus the installation of a new German-made power pack, the MTU diesel engine 6V 183TC 22. This develops 220kw and features a new transmission made by the German company ZF. In addition, the cooling system was modified and a new air intake filter added.

The first vehicles were modified under the NDV program in 1996. Along with the NDV modification, but not part of the program, a new track was issued. The Diehl 513 track has a longer in-service life than the earlier Diehl 213B track. The *Bundeswehr* operates the M113 family of vehicles in eighteen main variants, some of which have sub-variants. While all early M113s and variants were modernized to the M113G and M113GA1 standard, the A2 modifications, as well as the NDV 1 and 2 programs, were applied only to selected variants. In the following list, all M113 main variants are given together with the German name, some additional information of sub-variants and upgrade programs:

Soldiers of Raketenartilleriebataillon *150 based in Wesel prepare a rocket during barracks training. The tail wings of the rocket are just about to be mounted. The* Lenkraketenwerfer Startfahrzeug *(LRakWfSF) Lance M667 ballistic missile launcher is based on the M548 Cargo Carrier. Its truck bed has been extended ending in a rear drop ramp. Note the crew cab folded away on the front left side of the vehicle. (Michael Jerchel)*

Side view of a Lade-Transport-Panzer *(LadeTrspPz) Lance M668 missile loader transporter of* Raketenartilleriebataillon *150. The vehicle, which is painted in the standard NATO three-color camouflage scheme, was seen in 1990. The missile loader transporter had a combat weight of 11.7 tons, including two missiles. Its crew consists of two soldiers. Like the launcher, the vehicle is powered by the 6 V 53 6-cylinder diesel engine that is also used in the M113A1 vehicles. (Michael Jerchel)*

The Beobachtungs-Panzer-Artillerie *(BeobPzArt) M113 is used by the forward observation officers of the German artillery. Development of the BeobPzArt M113GA1 started in 1973, and the first vehicles entered service in 1981. The* Bundeswehr *is the only user of this M113 variant, and a total of 320 vehicles were issued. The picture shows a BeobPzArt M113GA1 of 3rd Battery,* Panzerartilleriebataillon *355 during an exercise in the early 1990's. The vehicle bears the standard NATO three-color camouflage and special exercise markings. (Yves Debay)*

- *Beobachtungs-Panzer-Artillerie* (BeobPzArt) M113, artillery observation post vehicle, A1 standard, A0 modification.
- *Mannschaftstransportwagen* (MTW) M113, armored personnel carrier, A1 and A2 standard, A0 modification, out of service since 1993.
- *Krankenkraftwagen* (KrKw) M113, armored ambulance, A1, A2 and NDV 1 standard, A0 modification.
- *Panzer-Mörser* (PzMrs) M113, mortar carrier, A1, A2 and NDV 2 standard, A0 modification.
- Schreib-Funk-Panzer (SchrFuPz) M 113, radio teletype vehicle used with VHF and HF radio equipment, A2 and NDV 1 standard, A0 modification.
- *Richt-Funk-Panzer Multiplex* (RiFuMuxPz) M113, radio-relay-access unit vehicle, A1 and A2 standard.
- *Gefechtsstands-Panzer* (GefStdPz) M577, command post vehicle, A1, A2 and NDV 2 standard.
- *Flieger-Leit-Panzer* (FlgLtPz) M113, forward air-controller vehicle, A1 and A2 standard, A0 modification.
- *Führungs-Funk-Panzer* (FueFuPz) M113, command/control vehicle, used in several variants including HF 400W, EVOX 1/3, EVOX 1/3A and EVOX 1/6, A1 and A2 version, A0 modification.
- *Feuerleit-Panzer Mörser* (FltPzMrs) M113, fire direction center mortar units, A1, A2 and NDV 1 standard, A0 modification.
- *Feuerleit-Panzer Artillerie* (FltPzArt) M113, fire direction center artillery units, A1, A2 and NDV 1 standard, A0 modification.
- *Geräteträger-Rechenverbund* (GerTrgRechnVbu ADLER) M113, digital data transmission cell, A1, A2 and NDV 2 standard, A0 modification.
- *Radarträger* RATAC (RadarTrgRATAC) M113, artillery-observation radar vehicle, A1, A2 and NDV 2 standard, A0 modification.
- *Radar Träger Artillerie Aufklärung* (RadarTrgArtAufkl) Green Archer M113, mortar location radar vehicle, A1 standard, out of service since the mid-1990s.
- *Lenkraketenwerfer Startfahrzeug* (LRakWfSF) Lance M667, ballistic missile launcher, 26 vehicles were in service since 1976, out of service since 1992.
- *Lade-Transport-Panzer* (LadeTrspPz) Lance M668, missile loader transporter, 26 vehicles were in service since 1976, out of service since 1992.
- *Minenwurf-Panzer* (MiWfPz) M548 Skorpion, mine scattering vehicle, A1 and NDV 1 standard.
- *Fahrschul-Panzer* (SchulPz) M113, driver training vehicle, A1, A2 and NDV 1 standard.

This Beobachtungs-Panzer-Artillerie *(BeobPzArt) M113GA1A0, which belongs to* Panzerartilleriebataillon *115, was seen during the last CMTC exercise conducted by the* Bundeswehr. *The BeobPzArt M113GA1A0 is fitted with a vehicle navigation system, a PERI D 11 double periscope, laser range finder, and a data processing unit. The PERI D 11 is connected with a laser range finder, and the whole system is called OZVA* (Optische Ziel Vermessungsanlage – *optical target measuring system). The crew enters the data into the DEA 64 processing unit from which data is transferred via a data radio link to the fire direction center of the artillery unit. In order to allow the extensive surveillance equipment to be mounted in the M113GA1, the roof of the BeobPzArt was modified and raised. Note the MILES sensor equipment attached to the side and rear of the vehicle.*

In the armored brigades of the Bundeswehr, *the* Beobachtungs-Panzer-Artillerie *(BeobPzArt) M113GA1A0 operates with the armored fighting units along the forward line of own troops. In the late 1980s, it became obvious that the BeoPzArt M113GA1A0 could not follow the Leopard 2- and Marder 1-equipped combat units. Therefore, it was decided that the vehicle would be replaced with a special version of the Leopard 1. However, due to funding problems this plan was abandoned in 2002. The picture shows a vehicle of* Panzerartilleriebataillon *115 during a CMTC exercise at the US training area at Hohenfels.*

Like most other M113 user nations, Germany uses the vehicle in the armored ambulance role. The pictured vehicle belongs to Panzerlehrbataillon *93 and is an M113GA2A0, which was photographed at the Munster training area in 1999. Note that the vehicle is fitted with EFTs. In addition to its three-man crew of driver, commander and medic, the vehicle can transport up to eight sitting wounded soldiers or four wounded soldiers on stretchers.*

Called Krankenkraftwagen *M113GA2A0, the armored ambulance is used in the German Army in armored infantry units, tank units, armored reconnaissance units, light infantry units, and tank killer units. Developed based on the MTW M113 armored personnel carrier, 650 vehicles were issued. The pictured vehicle belongs to* Panzerbrigade *12 and was seen during Exercise "Hessischer Löwe 2000."*

The Panzer-Mörser (*PzMrs*) M113GA1 mortar carrier of the German army is fitted with a 120mm Tampella mortar. From 1968 onwards, Thyssen Henschel in Kassel transformed 500 M113A1s into mortar carriers. The vehicle is used by the mortar platoons of the famous Panzergrenadier armored infantry units, as well as in mountain and light infantry units. Note the base plate attached to the rear of the vehicle. It is possible to dismount the mortar from the vehicle. The later A2 version of the PzMrs M113G, which is also fitted with EFTs, does not have this feature.

This picture shows a Schreib-Funk-Panzer (*SchrFuPz*) M113GA1 belonging to the German-French Brigade during an exercise in southern France in 1995. Note the 40W ASB antenna at the rear of the vehicle. While the vehicle is also operational when moving, its fully operational range can only be reached when operated stationary with the large antenna.

This Panzer-Mörser (PzMrs) M113GA2 fitted with EFTs, which belongs to Gebirgsjägerbataillon 571, is shown during a live-firing exercise in 1998. The loader of the five-man crew of the vehicle is just dropping a 120mm round into the mortar tube. Note the storage rack on the side of the vehicle in which poles for the camouflage net are stored.

Panzer-Mörser *(PzMrs) M113GA2s of* Gebirgsjägerbataillon 571 *demonstrate their cross-country capability during a live-firing exercise at the Schwarzenborn training area in 1998. The headlights of the first vehicle are quite visible, a good way to distinguish German M113s. The headlights consist of the main driving light, indicators and a dimmed light for tactical night driving.*

This photo of a Panzer-Mörser *(PzMrs) M113GA2 of* Gebirgsjägerbataillon *571 [Mountain Infantry Battalion] provides a good view of the EFTs. The external fuel tanks can hold up to 360 liters (95 gallons) of diesel fuel. The EFTs offer the same ballistic protection as the vehicle's hull. By moving the fuel to the outside of the vehicle, the fire hazard inside the crew compartment was reduced and additional storage room was gained.*

The German mortar carrier variant is equipped with a 120mm Tampella mortar. The weapon has a combat range of between 450 and 6350 meters (492 and 6940 yards). Some 63 rounds are carried inside the Panzer-Mörser *(PzMrs) M113GA2. Available ammunition consists of HE, smoke/marker and illumination rounds. Here the mortar crew adjusts the mortar using the PERI-R 16A1 aiming device. The crew of the mortar carrier consists of five soldiers, and the vehicle has a combat weight of 11.8 tons.*

Here a PzMrs M113GA2 fitted with EFTs belonging to the School of Infantry in Hammelburg, Bavaria, takes part in a live-firing demonstration in 1999. Clearly visible is the bank of smoke dischargers mounted on the top of the front glacier plate of nearly all of the German M113 variants.

This Minenwurf-Panzer M548GA1 Skorpion, which belongs to Panzerpioniercompanie 550 of the German-French Brigade, was seen during a exercise in southern France in 1995. The mine scattering system developed and produced by Dynamit Nobel is mounted on the M548GA1 chassis. It consists of six mine scattering units. The total weight, including 600 mines, is 4 tons, and the combat weight of the complete vehicle is 12 tons. In addition to the mine scattering system, the Skorpion is fitted with a 7.62mm MG3 machine gun. The crew of the vehicle consists of a driver and commander. Mounted in the crew cab of the vehicle is the EPAG (Einstell-, Prüf- und Abfeuer-Gerät). With the EPAG, the mine scattering is controlled and can be self-tested.

The German *Bundeswehr* operates 301 *Minenwurf-Panzer* M548GA1 Skorpion mine scattering systems within their armored engineer units. Belonging to *Panzerpioniercompanie* 550 of the German-French Brigade, this *Minenwurf-Panzer* M548GA1 Skorpion was seen during an exercise in southern France in 1995. The mine scattering system consists of six mine scattering units, each of which contains 100 mines.

The Minenwurf-Panzer M548GA1 Skorpion mine scattering system is fitted with six mine scattering systems, each of which contains 100 AT-2 mines in five magazines of 20 mines each. Using the mine throwing system, a minefield of 1500m x 200m (1640 yards x 219 yards) can be laid within 10 minutes. The operational time of the mines can be preset ranging between 3 and 96 hours. Once the preset operational time has been reached, the mine destroys itself. Part of the NDV 1 program, the M548A1 Skorpion vehicles received new tracks and track guards. The pictured vehicle, which is one of the upgraded vehicles, belongs to Pionierbataillon *4 based in Bogen in southern Germany.*

A total of 301 Minenwurf-Panzer *M548GA1 Skorpion mine scattering systems are in service with the* Bundeswehr. *The vehicle entered service in 1986 and was issued to the armored engineer companies on brigade level, as well as to the engineer battalions on division level. This M548A1 Skorpion was photographed in the late 1990s during a CMTC exercise of the* Panzerbrigade *12. It belongs to the armored engineer company of the brigade,* Panzerpionierkompanie *120.*

This pair of Gefechtsstands-Panzer *(GefStdPz) M577GA2s belongs to the HQ of the German-French Brigade. They were photographed during an exercise in southern France in 1995. Note the antenna masts erected on the vehicles and the generators in operational configuration. The A2 conversion of the German GefStdPz M577GA1 was carried out in the late eighties. In the* Bundeswehr, *the vehicle is also known as the GefStdPz M113GA2.*

The Gefechtsstands-Panzer *(GefStdPz) M577GA2 command post vehicle is used in the* Bundeswehr *in four different roles: command cell, information cell, teletype cell, and air liaison cell. The M577GA2 was developed in the* Bundeswehr *based on the earlier M577G and M577GA1. The most prominent distinguishing feature of the A2 version is the generator, which has been moved from the center slightly to the right. This vehicle belonged to the German KFOR contingent and was seen in Prisren in 2002.*

The Feuerleit-Panzer Mörser *(FltPzMrs) M113GA2A0 is used as a fire-direction center vehicle in the mortar elements of the armored infantry units. Note the storage basket that was mounted on all German M113 variants instead of the trim vane after the German M113 vehicle family lost their amphibious capability due to increased weight in the early 1990s. Also note the latest Diehl 512 track.*

Inside view of the rear compartment of a Feuerleit-Panzer Mörser *(FltPzMrs) M113GA2A0, fire-direction center mortar units. Note the map board, GPS navigation system and DVA-M mortar-fire data computer.*

A Geräteträger-Rechenverbund *(GerTrgRechnVbu ADLER) M113GA2 NDV 2 belonging to* Panzerartilleriebataillon *2 of* Panzerbrigade *14 crosses an M2 amphibious bridge during Exercise "Hessischer Löwe 2002." The vehicle-mounted ADLER computer system* (Artillerie-Daten-Lage und Einsatz-Rechnerverbund) *basically is a data processing system that links command elements with fire units. It also acts as an interface to the communication and data transmission systems of other NATO states.*

These two Führungs-Funk-Panzer *(FueFuPz) M113GA2A0 NDV 1 vehicles belong to* Panzerbataillon *84 and were seen during an exercise in 2001. The vehicles are fitted with EFTs, new track guards and the new Diehl 512 track. Both vehicles have the 40W ASB* (Antenne Standbetrieb) *erected and their radio sets include three SEM 90 radios per vehicle.*

Rear view of the same Geräteträger-Rechenverbund *(GerTrgRechnVbu ADLER) M113GA2 NDV 2 of* Panzerartilleriebataillon *2. Visible are the EFTs, modified track guards, the auxiliary power unit mounted on the roof to the right in the rear of the vehicle, and the antenna mast mounted on the left rear side of the roof.*

This picture of a Führungs-Funk-Panzer *(FueFuPz) M113GA1 of* Panzerjägerkompanie *340 shows the vehicle's rear and roof. Clearly visible are the 6-meter (20-foot) 40W ASB antenna mast, generator box and storage box for additional radio equipment. (Michael Jerchel)*

A Führungs-Funk-Panzer *(FueFuPz) M113GA2A0 of a* Panzergrenadier *armored infantry unit crosses an M3 amphibious bridge during a large exercise in early 2002.*

This picture shows a Führungs-Funk-Panzer *(FueFuPz) M113GA2A0 belonging to* Panzerbrigade *14 during Exercise "Hessischer Löwe 2002." The vehicle differs from the* Schreib-Funk-Panzer *(SchrFuPz) M113 only in its radio equipment. The German Army uses more than 18 known variants.*

This picture shows a Führungs-Funk-Panzer *(FueFuPz) M113GA2A0 belonging to* Panzerbrigade *14. The FueFuPz M113GA2A0 has a crew of four that consists of the driver, commander and two radio operators. For self-defense purposes the vehicle is fitted with a 7.62mm MG3 machine gun.*

A FueFuPz M113GA2A0 command/control vehicle leads a column of Panzerhaubitze 2000 self-propelled howitzers into a new fire position during Exercise "Hessischer Löwe 2002." The vehicle belongs to Panzerartilleriebataillon 2 of Panzerbrigade 14.

A Panzerhaubitze 2000 self-propelled howitzer can be seen in the background in this photograph. The other two vehicles in the picture are different variants of the M113 family of vehicles in service with the German Army's artillery. In the foreground is a Führungs-Funk-Panzer (FüeFuPz) M113GA2A0, and behind it is a Geräteträger-Rechenverbund (GerTrgRechnVbu ADLER) M113GA2 NDV 2. Note that both vehicles are fitted with the Diehl 512 track, which was issued in the late 1990s. Also notice that the NDV vehicle has new track guards.

For driver training, the Bundeswehr *operates a fleet of* Schulpanzer *M113GA1s and M113GA2s. Pictured here is the* Schulpanzer *M113GA1, which is only equipped with weather protection. The newer M113GA2 is fitted with a cabin for the driving teacher and additional students; it is similar to the one used on the Leopard 1 driver training vehicle. (Michael Jerchel)*

This Radarträger *RATAC (RadarTrgRATAC) M113GA1 artillery observation radar vehicle, which belongs to* Panzerartilleriebataillon *35, was seen in 1990. The radar, the RATAC-S, has a detection range of 24km (15 miles). The vehicle has a combat weight of 13kg (29 lb.) and the radar is erected on a 7-meter (23-foot) mast. The vehicle, which entered service in 1978, has a crew of four soldiers. (Michael Jerchel)*

A total of 73 Radarträger *RATAC (RadarTrgRATAC) M113GA1 vehicles were fitted with the RATAC-S radar. In order to allow storage of the radar and the mast system on the roof of the vehicles, the large box-shape modification was built. (Michael Jerchel)*

Known as the M113GA2 ABRA (Artillerie Beobachtungs Radar Anlage), *the pictured vehicle replaced the* Radarträger *RATAC (RadarTrgRATAC) M113GA1 artillery observation radar vehicle. The main difference between modern M113GA2 ABRAs and the older RadarTrgRATAC M113GA1 is the new DR-PC1A RATAC* (Radar de Tir pour Artillerie de Campagne) *battlefield radar with enhanced performance. The radar, which is co-produced by France and Germany, works in the 300-600Mhz band and can detect enemy forces up to a range of 38km (24 miles). The RATAC radar is mounted on a 7-meter (23-foot) mast and has a7kW HF performance.*

Danish M113

During the 1990s, Denmark operated a fleet of 738 vehicles of the M113 family. The first batch, which was delivered between 1962 and 1964, consisted of M113s powered by petrol engines. In the Danish Army, the M113 replaced obsolete American half-tracks dating from World War II. Beginning in 1967, the Danish Army received a second batch of M113s, this time of the A1 standard. While this delivery was still in progress, the M113s already in service were upgraded to the M113A1 standard in Denmark. Beginning in 1978, a third batch of M113A1s was delivered to the Danish Army. Of the 738 M113A1s in service with the Danish Army, most are in the role of an armored personnel carrier, while 56 are mortar carriers. This includes M106A1s armed with the 107mm mortar and M125A1s armed with the 81mm mortar. Another 56 vehicles are fitted with the TOW anti-tank weapon system. While eight M113A1s are fitted with the Green Archer Mortar Location Radar, another 32 field the M/74 ZB 298 radar. Also used in Denmark is the M579A1 fitters vehicle, 20 of which were issued. Instead of the M577, which did not enter the Danish Army inventory, the M113A1 with a roof-mounted generator is used as a command post vehicle. The Danish Army uses the M113A1 as an armored ambulance and engineer section vehicle.

A German landing craft brings Danish M113A1s ashore during an amphibious exercise in Denmark in the early 1990s. Note that the vehicles bear red OPFOR markings. On Danish M113A1a, the machine gun mount near the commander's cupola is fitted either with a 12.7mm M2 HB machine gun or the German 7.62mm MG3 machine gun. (Yves Debay)

At the time of writing, 257 vehicles of the Danish M113A1 fleet have been upgraded to M113 G3 DK. A similar upgrade is carried out for the German *Bundeswehr* under the NDV 2 program. The Danish upgrade program for the M113 G3 DK consists of the MTU 6V 183 TC22 EURO II engine, ZF LSG 1000 transmission, a new braking system, improved suspension, EFTs (External Fuel Tank), and Diehl 513 tracks. *Flensburger Fahrzeugbau Gesellschaft mbH* performs the upgrade in Germany. While the M113A1 was still the most important armored personnel carrier for Denmark's mechanized infantry in the 1990s, in the first years of the new millennium the Piranha III entered service in order to replace the M113s on peace-support operations. A version of the M113 used only by the Danish Army is the M113A2 fitted with a 25mm machine cannon turret; the vehicle is also known as the M92. The M92 was developed and built in Denmark by E. Falck Schmidt in Odense in the early 1990s. The Danish Army purchased 50 M113A2 chassis in the U.S. and then fitted them with the turret in Denmark.

The Danish Army uses the standard M113A1 as an armored command post instead of the M577. Here such an M113A1 command post vehicle can be seen during an amphibious landing operation. The Danish command version of the M113A1 is recognizable by the large box containing a generator mounted at the rear of the vehicle. (Yves Debay)

During Exercise "Action Express 93," Danish armored units acted as aggressors for the AMF(L). This M125A1 with an 81mm mortar belongs to an armored infantry unit. The Danish Army operates 56 mortar carriers. While the bulk of these are 81mm-mortar-equipped M125A1s, some are earlier 107mm-mortar-equipped M106 versions. Note the "Devil" graffiti and the flag.

In the 1990s, the M113A1 was still the standard APC of the Danish Army. Here a pair of well-camouflaged M113A1s participates in Exercise "Action Express 93" in Denmark. In the armored battle groups, the M113A1 worked alongside the Leopard 1 main battle tank and the M41DK reconnaissance light tank. Despite the introduction of the Piranha III wheeled APC, the M113 family of vehicles will stay in service with the Danish Army for several more years.

The Danish Army also employs the M113A1 as a TOW anti-tank missile carrier. There are 56 such vehicles in service, and the vehicle replaced the Land Rover 88 Lightweights used earlier in the role. The TOW launcher is mounted in front of the large cargo hatch. Ten spare missiles can be carried inside the vehicle. A .50-caliber M2 HB machine gun serves as secondary armament. This picture was taken during the AMF(L) exercise "Action Express 93" in Denmark.

A curious version of the M113A1 is the simulator main battle tank. In order to minimize exercise damage in Denmark, M113A1s were used instead of the much heavier Leopard 1 during the early '90s. The vehicles were fitted with a mock turret in order to make them appear more like a main battle tank. This vehicle was seen during Exercise "Action Express 93" on Sealand Island. (Yves Debay)

The first time Danish M113A1s fitted with passive add-on armor were in Bosnia in late 1992. The commander's cupola was fitted with the armored weapon station and machine gun shield already known from the ACAV. Note that in addition to the add-on armor, the vehicle is fitted with two banks of four smoke grenade dischargers on the left and right side.

Like the armored personnel carriers, the M113A1 ambulances of the Danish SFOR contingent were also fitted with passive add-on armor. Note that even the ACAV-type machine gun shield and armor-plated commander's weapon station were fitted to the vehicle. This M113A1 ambulance, nicknamed "Jolly Jumper," was seen in March 1997 near Doboj in the sector of the NorthPol Brigade.

Like several other nations, Denmark operates the M113A1 as an armored ambulance. The vehicle can evacuate from the battlefield four wounded soldiers on stretchers and two sitting wounded soldiers. The vehicle does not carry any armament but it does have a mounting for the .50-caliber machine gun. (Yves Debay)

This vehicle, which belonged to the Danish UNPROFOR contingent in Bosnia, was photographed near Kiseljak in December 1992. Note the red and white Danish national badge and the standard UN color and markings, as well as the UNPROFOR number plate.

Like UNPROFOR and IFOR, the Danish troops of the SFOR contingents used the M113A1 with add-on armor as armored personnel carriers for patrolling. Some of the vehicles were fitted with a special antenna mast at the rear. This became necessary because radio communication during patrols in the mountainous countryside of Bosnia was very difficult. The mast, which could be raised several meters, partly solved the problem. This picture was taken in the NorthPol Brigade sector in 1998.

This M113A1 with add-on armor, fitted with a .50-caliber machine gun, which belongs to the Danish KFOR contingent, was seen in Mitrovica in March 2000. Note the additional modifications made by the crew in order to protect themselves against stones thrown by rioters, as well as the razor wire placed on the hull to keep people from climbing on the vehicle.

A column of M113A1s of the Danish KFOR contingent awaits orders during the March riots in Mitrovica in 2000. All vehicles are fitted with passive add-on armor kits. Note the anti-riot equipment stored on the vehicles.

The M92 saw its first operational service with the Danish IFOR troops in 1995 and 1996. This M92 was seen in the summer of 1996 outside the NorthPol Brigade HQ Camp in Doboj. For the deployment to former Yugoslavia, six M92s were fitted with a passive add-on armor package. The armor was fitted to the vehicle's front, sides and turret. Inside the vehicle, spall liners were mounted. The add-on armor can stop a conventional 14.5mm round.

This M92, which belongs to the Danish KFOR contingent, was photographed near Mitrovica at a mobile vehicle checkpoint in March 2000. The M92 is an M113A2 chassis fitted with an Oto Melara machine cannon turret that is armed with a 25mm Oerlikon-Contraves machine cannon and a coaxial 7.62mm MG3 machine gun. For Peace Support Operation deployment, the vehicles were fitted with a passive add-on armor, and safety foam balls were used in the EFTs.

An M92 of the Danish KFOR contingent guards a mobile vehicle checkpoint west of Mitrovica in March 2000. The M92 is an M113A2 chassis fitted with an Oto Melara machine cannon turret. The two-man turret is armed with a 25mm Oerlikon-Contraves machine cannon and a coaxial 7.62mm MG3 machine gun. The sighting equipment of the turret, which is made by Zeiss in Germany, features a day-and-night sight as well as a thermal imaging sight.

Danish M92 in position at the important Doboj bridge. With the 25mm Oerlikon-Contraves KBA machine cannon the vehicle is able to engage an enemy up to a range of 1500 meters (1639.5 yards), while the coaxial 7.62mm has a combat range of up to 1200 meters (1312 yards). The add-on armor on the vehicle's side is clearly visible. While the ceramic-and-steel add-on armor on the turret was produced in Denmark, United Defense LP supplied the add-on armor of the chassis.

The Danish M92s, which are based on the M113A2 chassis, were fitted with EFTs. This created more space in the vehicle's troop compartment and for the installation of the Oto Melara two-man turret. For the IFOR deployment, the vehicles were fitted with passive add-on armor, and safety-foam balls were used in the EFTs.

A pair of Danish M113s take part in Exercise "Strong Resolve 2002." The vehicle in the foreground is an M113 G3 DK TOW anti-tank missile carrier, and an M92 is in the background. Note that add-on armor is mounted on the vehicle in the front; it can be easily identified by the greased mounting points standing out against the side of the vehicle

A Danish M113 G3 DK of the Danish Reaction Brigade during Exercise "Strong Resolve 2002" in Poland. The upgrade package included the new MTU 6V 183 TC22 EURO II, EFTs, ZF LSG 1000 transmission, improved suspension, Diehl 513 tracks, new Stewart-Warner type 10560 M 200 AMP generator, and improved steering and braking systems. Other modifications were a front-mounted bank of six smoke grenade dischargers, modified headlights, and an add-on armor package.

Norwegian M113

The Norwegian Army uses the M577A1 command post vehicle in the HQ element of their armored and artillery units. Like other nations, Norway uses the M577A1 with a tent set up at its rear in order to provide more working space for the HQ staff.

The Norwegian Army has operated several variants of the M113 vehicle family since 1977. In total, the Norwegian M113 fleet consists of approximately 670 vehicles. Most of these vehicles were received directly from the USA, but some were bought used from Germany and the Netherlands. The bulk of the Norwegian M113 fleet is made up of A1-version vehicles, and in 1997 the army upgraded their M113s with a modified suspension, cooling system and final drive. The conversion kits were bought in Canada from the Canadian Commercial Corporation. Other M113 variants used by the Norwegian Army include the ambulance, basic personal carrier, 81mm mortar carrier, and command post vehicle. The Norwegian Army deployed several M113A1s fitted with passive add-on armor to Bosnia in 1996 with its IFOR contingent.

Norway has built two own variants on M113 chassis: the NM 135 Mechanized Infantry Combat Vehicle (M113A1 with 20mm turret) and the NM 142 *Rakettpanserjager* (M113A1 fitted with Kvaerner Eureka TUA turret). While the fleet of NM 135s is only 53 vehicles strong, 97 NM 142s were built. In addition to its M113 fleet, the Norwegian Army operates 86 M548A1 tracked cargo carriers. The bulk of these vehicles is used as ammunition supply vehicles for the artillery units equipped with the M109 self-propelled howitzer. The M548A1 is also used as a workshop vehicle in maintenance units.

Like the armies of countless other user nations, the Norwegian Army employs the M113A1 in the armored ambulance role. Inside the vehicle, there is enough space for medical personnel and four wounded soldiers on stretchers. The vehicle, which belongs to Brigade "North," was seen during the AMF(L) exercise "Arctic Express 94."

Among its M113 vehicle fleet, the Norwegian Army operates 86 M113-based M548A1 tracked cargo carriers. The vehicles are mainly used to support the M109A3-equipped artillery batteries. This M548A1 is fitted with a hydraulic crane for handling pallets.

In addition to its role of ammunition-supply vehicle, the M548A1 is also used in maintenance units. Here an M548A1 assists an M113A1 crew that is suffering from engine problems. Norway also developed a recovery vehicle called the NM 84 that is based on the M548A1. It features a 5-ton crane and an 18-ton winch. (Yves Debay)

A forward observation officer from a mortar platoon uses a Norwegian M113A1. The vehicle was seen in northern Norway during the NATO exercise "Strong Resolve 98."

Within the Norwegian Army, the M113A1 is known as the NM 113. Here a NM 113 in its basic configuration is used by the HQ element of the 5th Infantry Brigade. Note that the driver and commander have windscreens fitted in front of their hatches to protect them against the ice-cold Norwegian winter wind. (Yves Debay)

Seen here are two examples of the 670 vehicle-strong Norwegian M113A1 fleet. The vehicle on the left is an NM 135 Mechanized Infantry Combat Vehicle, while the one on the right is a basic M113A1 Armored Personnel Carrier fitted with a .50-caliber machine gun.

One variant of the M113A1 used in the Norwegian Army is the M125A1 81mm mortar carrier. An 81mm mortar is mounted inside the vehicle on turntable that can be rotated 360°. Clearly visible is the three-part roof hatch that covers the large opening in the roof over the rear compartment in which the mortar is situated.

The NM 135 is fitted with a one-man turret built by Hägglunds in Sweden. A 7.62mm MG3 machine gun is mounted coaxially with the cannon on the right side of the turret. In addition, a bank of three smoke grenade dischargers is mounted on either side of the turret. (Yves Debay)

The Norwegian NM 135s are based on the M113A1 APCs that were delivered to the country in 1977. Though the vehicle's one-man turret was built by Hägglunds Vehicles in Sweden, the conversion from the M113A1 to the NM 135 and the mounting of the turrets was carried out in Norway.

The NM 135 is used in the Norwegian Army's "Storm" infantry units. These units are basically the equivalent of armored infantry units in the armies of other nations. Norway fitted 53 M113A1s with a one-man turret featuring a 20mm Rheinmetall Rh 202 cannon; the variant was named NM 135. In 1997, the Norwegian Army started to replace the NM 135 with the CV 9030N armored infantry fighting vehicle. Here a Storm unit takes part in the action during the NATO exercise "Arctic Express 94," which was held in northern Norway. The vehicles belong to the Brigade North, which forms Norway's most important armored unit.

The Hägglunds one-man turret was mounted on the NM 135 behind the engine compartment. The main armament of the turret is a 20mm Rheinmetall Rh 202 cannon with a coaxial 7.62mm MG3 mounted on the turret roof. With the turret mounted behind the engine compartment, the commander's cupola was moved directly behind the driver's hatch.

Rear view of an NM 135 of the Norwegian Army. Visible are the one-man 20mm Hägglund turret and the external storage boxes mounted on the rear of the vehicle, as well as the power-operated ramp, which is seen here during closing procedures after the infantrymen have returned to their vehicle.

The NM 142 Rakettpanserjager *is an M113A1 fitted with a TOW Armored Launching Unit (ALT). The vehicle has a crew of four soldiers consisting of a driver, commander, TOW operator, and loader. In addition to the two TOW launchers in the ALT, the vehicle is fitted with a 7.62mm MG3 machine gun and six smoke grenade dischargers. All types of TOW missiles (including the 2B version) can be used with the launchers of the ALT. With the ALT, the NM 142 has a combat weight of 14 tons.*

The TOW Armored Launching Unit (ALT) of the NM 142 Rakettpanserjager *was designed and produced by the Norwegian company Kvaerner Eureka. The one-man turret can be traversed 360°, has a weight of 930kg (2050 lb.) and features two TOW launching tubes, as well as the sighting equipment for the anti-tank weapon system. In order to mount the turret in the center of the M113A1 chassis, the commander's copula was moved to the right side.*

Here a Norwegian NM 142 Rakettpanserjager *patrols west of Pristina in Kosovo. The vehicle belongs to the Telemark Battalion, which formed the Norwegian KFOR contingent in 1999. The vehicle has a crew of four, but only the driver and commander are visible. In addition to the two ALT TOW turrets, the vehicle is fitted with a 7.62mm MG3 machine gun and six smoke grenade dischargers.*

While the bulk of the vehicles deployed to Kosovo under KFOR with the "Telemark Battalion" in 1999 were Patria XA 186 6x6 wheeled APCs, the deployed engineer asset used a fleet of different M113 variants. Here a Norwegian M113A1 passes through Kosovo Polje. Note that the vehicle is painted in plain green and does not wear the typical Norwegian three-color camouflage of green, brown and black.

Taken in March 2000, this photograph shows the rear of one of the NM 142s deployed to Kosovo with the Norwegian KFOR contingent. Clearly visible are the six smoke grenade dischargers situated on the rear of the ALT. Note also the modified cargo hatch over the troop compartment, in which an M17 periscope is mounted. The loader manually loads the TOW launcher. For reloading, the cargo hatch is opened and then the loader lifts the missile container into the launcher by hand.

This Norwegian Army M113 was seen in March 2000 during a mine clearing operation at the Kacanik Pass. The vehicle is fitted with a Pearson mine roller. Bolted-on passive add-on armor has been fitted to the vehicle's hull. The vehicle armament consists of a pintle-mounted .50-caliber M2HB heavy machine gun. Note the shield that is mounted in order to protect the commander when operating the machine gun. The vehicle, which is one of the M113s upgraded by the Canadian Commercial Corporation, features EFTs.

End of March 2002: This Norwegian M113 belonging to an ISAF EOD team was seen on Bagram Airport in Afghanistan. The vehicle is fitted with a Pearson mine roller. The vehicle is fitted with bolted-on passive add-on armor and features a special weapon station for a .50-caliber sniper rifle. Able to be traversed 360°, the system uses a TV camera to engage unexploded ammunition and mines. The turret is mounted to the right of the commander's cupola. Note the wire cutters and the sand-and-brown camouflage pattern.

More M113 User Nations

The Portuguese army operates a fleet of nearly 550 vehicles of the M113 family. This includes the M113A1 APC, the M106A1 and M106A2, the M125A1 mortar carriers, the M577A1 and M577A2 command post vehicle, the M113A1 with roof-mounted TOW, the M901 ITV, and the M 730A1 Chaparral surface-to-air missile launcher. The Portuguese Army received the first vehicles in 1978, and from 1996 onwards, the M113A2 entered service. The pictured M113A1 APC was seen guarding the MNB(W) HQ in Pec in Kosovo in early 2000. The vehicle belongs to the Portuguese KFOR contingent that formed part of the Italian-led Multinational Brigade West.

Like other Arabian states, Saudi Arabia operates the M113 in large quantities and countless variants. An estimated 2000 US- and Italian-production vehicles are in service. At the time of writing, Saudi Arabia is upgrading its fleet of M113A1- and M113A2-family vehicles to the A3 standard. The picture shows an M577A2 command post vehicle during the build-up to the 1991 Gulf War. (Yves Debay)

Belonging to the Kuwaiti free brigade, this M577A1 is one of several vehicles that managed to escape Iraqi forces during the invasion of Kuwait in 1990. The vehicle subsequently took part in the 1991 Gulf War. (Yves Debay)

Egypt operates a fleet of more than 2000 M113s, most of which are of the A2 standard. The picture shows an M113A2 APC of an infantry unit fitted with EFTs. Other vehicles of the M113 family operated by Egypt are the M577A2 command post vehicle, the M548A1 ammunition supply vehicle, the M730 Chaparral air defence system, the M806A1 armored recovery vehicle, M106 A1 mortar carrier, and the M901 ITV. Exclusively used by Egypt is a version of the M577A2 fitted with a TRACKSTAR radar system and M113-based air defence vehicles fitted with a one-man turret and armed with two 23mm cannons and the Sakr Eye SAM. (Yves Debay)

This M113A1, which belongs to a Kuwaiti armored infantry unit, was seen during the 1991 Gulf War somewhere in the desert. The fleet of Kuwaiti M113s consists of approximately 108 vehicles. Variants in use are the M113A1 APC, the M557A1 command post vehicle and the M901A1 ITV, but the Iraqi forces destroyed or captured some of these during the 1991 Gulf War. In 1993, the Kuwaiti government ordered several M113A3 variants. (Yves Debay)